YOUR KNOWLEDGE HAS VALUE

- We will publish your bachelor's and
  master's thesis, essays and papers

- Your own eBook and book -
  sold worldwide in all relevant shops

- Earn money with each sale

Upload your text at www.GRIN.com
and publish for free

S. Narendar

Non-classical Wave Dynamics of Ultrathin Structures

GRIN Verlag

Bibliografische Information der Deutschen Nationalbibliothek:

Die Deutsche Bibliothek verzeichnet diese Publikation in der Deutschen National-
bibliografie; detaillierte bibliografische Daten sind im Internet über http://dnb.d-
nb.de/ abrufbar.

Dieses Werk sowie alle darin enthaltenen einzelnen Beiträge und Abbildungen
sind urheberrechtlich geschützt. Jede Verwertung, die nicht ausdrücklich vom
Urheberrechtsschutz zugelassen ist, bedarf der vorherigen Zustimmung des Verla-
ges. Das gilt insbesondere für Vervielfältigungen, Bearbeitungen, Übersetzungen,
Mikroverfilmungen, Auswertungen durch Datenbanken und für die Einspeicherung
und Verarbeitung in elektronische Systeme. Alle Rechte, auch die des auszugsweisen
Nachdrucks, der fotomechanischen Wiedergabe (einschließlich Mikrokopie) sowie
der Auswertung durch Datenbanken oder ähnliche Einrichtungen, vorbehalten.

Imprint:

Copyright © 2012 GRIN Verlag GmbH
Druck und Bindung: Books on Demand GmbH, Norderstedt Germany
ISBN: 978-3-656-29533-4

This book at GRIN:

http://www.grin.com/en/e-book/201982/non-classical-wave-dynamics-of-ultrathin-
structures

GRIN - Your knowledge has value

Der GRIN Verlag publiziert seit 1998 wissenschaftliche Arbeiten von Studenten, Hochschullehrern und anderen Akademikern als eBook und gedrucktes Buch. Die Verlagswebsite www.grin.com ist die ideale Plattform zur Veröffentlichung von Hausarbeiten, Abschlussarbeiten, wissenschaftlichen Aufsätzen, Dissertationen und Fachbüchern.

Visit us on the internet:

http://www.grin.com/

http://www.facebook.com/grincom

http://www.twitter.com/grin_com

Non-classical Wave Dynamics of Ultrathin Structures

S. Narendar

Defence Research and Development Laboratory, Kanchanbagh, Hyderabad-500 058

Abstract:

In this paper, the nonlocal elasticity theory has been incorporated into classical 1D-rod model to capture unique features of the rod like structures at Nanoscale, which are considered as ultra-thin structures, under the umbrella of continuum mechanics theory. The strong effect of the nanoscale has been obtained which leads to substantially different wave behaviors of nanoscale-rods from those of macroscopic rods. Nonlocal bar model is developed for nanorods. The analysis shows that the wave characteristics are highly over estimated by the classical rod model, which ignores the effect of small-length scale. The studies also show that the nonlocal scale parameter introduces certain band gap region in axial wave mode where no wave propagation occurs. This is manifested in the spectrum cures as the region where the wavenumber tends to infinite (or wave speed tends to zero). These results are also compared with the Born-Karman model and also with the second and fourth order strain gradient models. The results can provide useful guidance for the study and design of the next generation of nanodevices that make use of the wave propagation properties of single-walled carbon nanotubes.

Keywords: Nonlocal stress gradient model, Nonlocal strain gradient model, Lateral inertia, Wavenumber, Phase speed, Dispersion, Group speed, Nanorod, Born-Karman Model.

1. Introduction

A ultra-thin structure is defined as a material system or object where at least one of the dimensions lies below 100 nm. Nanostructures can be classified into three different categories: zero-dimensional; one-dimensional; two-dimensional. Zero-dimensional nanostructures are materials in which all three dimensions are at the nanoscale. A good example of these materials are buckminster fullerenes [1] and quantum dots [2]. One-dimensional nanostructures are materials that have two physical dimensions in the nanometer range while the third dimension can be large, such as in the carbon nanotube [3]. 2D nanostructures, or thin films, only have one dimension in the nanometer range and are used readily in the processing of complimentary metal-oxide semiconductor transistors [4] and micro-electro-mechanical systems (MEMS) [5]. Since the focus of this work is on one-dimensional nanostructures, all others from this point forward will cease to be discussed. One-dimensional nanostructures (here *nanorods*) have stimulated a great deal of interest due to their importance in fundamental scientific research and potential technological applications in nano-electronic, nano-optoelectronic and nano-electro-mechanical systems.

Classical continuum theories assume that the stresses in a material point depend only on the first-order derivative of the displacements, i.e. on the strains, and not on higher-order displacement derivatives. As a consequence of this limitation on the kinematic field, a classical continuum is not always capable of adequately describing heterogeneous phenomena. For instance, unrealistic singularities in the stress and/or strain field may occur nearby imperfections. Furthermore, severe problems in the simulation of localization phenomena with classical continua have been encountered, such as loss of well-posedness in the mathematical description. To avoid these types of deficiencies, it has been proposed to include higher-order strain gradients into the constitutive equations, so that the defects of the classical continuum may be successfully overcome [6-9]. The second-order strain gradients that are normally used for these purposes introduce accessory material parameters that reflect the microstructural properties of the material.

The higher-order gradients can improve the performance of the classical continuum in the sense that the dispersive behavior of the discrete model is reproduced with a higher accuracy, [10-12]. This is a direct consequence of the procedure that is commonly used to enhance the classical continuum with higher-order gradients: homogenization of the discrete medium may lead to higher-order gradients in a direct and straightforward manner. If regularization of singularities or discontinuities is required, higher-order gradients are used for smoothing the non-uniformity or singularities in the strain field. On the other hand, if a more accurate representation of the discrete microstructure is desired, the higher-order gradients are used to introduce a non-uniformity in the strain field.

Actually, the length scales associated with nanostructures like nanorods are such that to apply any classical continuum techniques, we need to consider the small length scales such as lattice spacing between individual atoms, surface properties, grain size. This makes a physically consistent classical continuum model formulation very challenging. The Eringen's nonlocal elasticity theory [13-15] is useful tool in treating phenomena whose origins lie in the regimes smaller than the classical continuum models. In this theory, the internal size or scale could be represented in the constitutive equations simply as material parameters. Such a nonlocal continuum mechanics has been widely accepted and has been applied to many problems including wave propagation, dislocation, crack problems, etc [16]. Recently, there has been great interest in the application of nonlocal continuum mechanics for modeling and analysis of nanostructures [17-27].

The study of wave propagation in nanostructures has attracted intensive attention in research because many crucial physical properties such as electrical conductance, optical transition and some dynamic behavior of carbon nanotubes (CNTs) are very sensitive to the presence of wave

The present wave propagation studies using nonlocal continuum model has shown that the wave behavior in a nanorod is drastically different compared to the behavior of local or classical model. Hence, the main objective of this paper is to bring out the main effects that the nonlocal scale parameter to the wave propagation in nanorods.In this paper, a nonlocal Euler-Bernoulli bar model is developed for analyzing the ultrasonic wave propagation in nanorods. The effect of nonlocal scaling parameter on the wave propagation in nanorods and also the variation of the escape frequency with e_0a is studied in detail. Here e_0a = 0.5 nm and 1.0 nm are used, where a = 0.142 nm (C-C bond length). Also Eringen's nonlocal elasticity theory is used to model the dispersion characteristics of ultrasonic waves in the nanorod. This nonlocal elasticity theory assumes that the stress at a reference point to be a functional of the strain field at every point in the body. It allows one to account for the small scale effect that becomes significant when dealing with micro and nanostructures. This theory is extended to obtain the nonlocal second and fourth order strain gradient models. At the end, the unstable second order strain gradient model is made stable by considering the lateral inertia effects along with the nonlocal stress gradient model in the nonlocal rod formulation.

2 Mathematical Formulation

2.1 A review on theory of nonlocal elasticity

According to the theory of nonlocal elasticity [13-15], the stress at a reference point $\mathbf{x}$ is considered to be a functional of the strain field at every point in the body. In the limit when the effects of strains at points other than $\mathbf{x}$ are neglected, one obtains local or classical theory of elasticity. The basic equations for linear, homogeneous, isotropic, nonlocal elastic solid with zero body force are given by:

$$\sigma_{ij,j} = 0, \sigma_{ij}(\mathbf{x}) = \int_V \alpha(|\mathbf{x}-\mathbf{x'}|,\xi)C_{ijkl}\varepsilon_{kl}(\mathbf{x'})dV(\mathbf{x'}) \, \forall \mathbf{x} \in V, \varepsilon_{ij} = 0.5(u_{i,j}+u_{j,i}),$$ where C_{ijkl} is the elastic modulus tensor of classical isotropic elasticity, σ_{ij} and ε_{ij} are stress and strain tensors respectively, and u_i is the displacement vector. $\alpha = \alpha(|\mathbf{x}-\mathbf{x'}|,\xi)$ is the nonlocal modulus or attenuation function incorporating the nonlocal effects into the constitutive equations. This nonlocal modulus is found by matching the curves of plane waves with those due to atomic lattice dynamics. Various different forms of $\alpha(|\mathbf{x}-\mathbf{x'}|)$ have been reported in [13]. $|\mathbf{x}-\mathbf{x'}|$ is the Euclidean distance, and $\xi = e_0 a/\ell$, where a is an internal characteristic length, e.g., length of $C-C$ bond (0.142 nm) in CNT, granular distance etc., and ℓ is an external characteristic length e.g., wavelength, crack length, size of the sample etc. e_0 is a nonlocal scaling parameter,

which has been assumed as a constant appropriate to each material in published literature and V is the region occupied by the body. On the other hand, parameter e_0 was given as 0.39 by Eringen [14].

As the constitutive equation of nonlocal elasticity involves spatial integrals which represent weighted averages of the contributions of the strain tensors of all points in the body to the stress tensor at the given point, it is difficult mathematically to get the solution of nonlocal elasticity problems. However, it is pointed out by Eringen [14] that the integral constitutive equation can be converted exactly into an equivalent differential form for some kernels. This, of course, provides a great deal of simplicity and convenience for the application of the theory of nonlocal elasticity. In what follows, the following form for the kernel function:

$\alpha(|\mathbf{x}|,\xi) = (1/2\pi\xi^2\ell^2)K_0\left(\sqrt{\mathbf{x}\times\mathbf{x}}/\xi\ell\right)$ will be adopted, which was suggested

by Eringen [14]. In this equation, K_0 is the modified Bessel function. Hooke's law for a uni-axial stress state by the nonlocal elasticity can be determined by [18] and given as $\sigma(x)-(e_0a)^2\sigma^{<ii>}(x) = E\varepsilon(x)$, where E is the Young's modulus of the material, x is the coordinate with the origin at the left end of one-dimensional structure. Eq. (5) is the uniaxial stress version of Eq. (3.19) of ref. [14].

2.2 Nonlocal strain gradient models

Expansion of the general integral constitutive equation of nonlocal elasticity (ref. [14], Eq. (3.17)) for $e_0a/L \ll 1$, retention of only the first two terms, and simplification to the case of uniaxial stress produces $\sigma(x) = E\left(\varepsilon(x)+(e_0a)^2\varepsilon^{<ii>}(x)\right)$. This is a second order strain gradient model with nonlocal scale effects.In the same way, retention of only the first three terms, and simplification to the case of uniaxial stress produces

$$\sigma(x) = E\left(\varepsilon(x)+(e_0a)^2\varepsilon^{<ii>}(x)+(e_0a)^4\varepsilon^{<iv>}(x)\right).$$

This is a fourth order strain gradient model with nonlocal scale effects. Next, we will derive the governing equation of motion for nanorod based on the above mentioned two constitutive relations asmentione above.

2.3   *Nonlocal governing partial differential equation for nanorods basedon nonlocal stress gradient model*

Fig. 1 schematically describes a nanorod under discussion and serves to introduce the axial coordinate x, the axial displacement $u = u(x,t)$, the length L, the Young's modulus E, and the density ρ. The displacement field and strain for this nanorod are given by $u = u(x,t)$ $\varepsilon_{xx} = u^{<i>}$. For thin rods nonlocal constitutive relation one dimensional form is $\sigma_{xx} - (e_0 a)^2 \sigma_{xx}^{<ii>} = Eu^{<i>}$, where E is the modulus of elasticity, σ_{xx} and ε_{xx} are the local stress and strain components in the x direction, respectively. The equation of motion for an axial rod can be obtained as $N^{<i>} = \rho A \ddot{u}$ where N is the axial force per unit length for local or classical elasticity and is defined by $N = \int_A \sigma_{xx} dA$.

Using these relations we have $N - (e_0 a)^2 N^{<ii>} = EAu^{<i>}$ Substitution of the first derivative of N in the above relation we obtain $N = EAu^{<i>} + (e_0 a)^2 \rho A \ddot{u}^{<i>}$. Substituting N into the equation of motion, we obtain

$$EAu^{<ii>} + (e_0 a)^2 \rho A \ddot{u}^{<ii>} = \rho A \ddot{u} \tag{1}$$

Eq. (1) is the consistent fundamental governing equation of motion for nonlocal rod model. When $e_0 a = 0$, it is reduced to the equation of classical rod model.

3 Ultrasonic Wave Characteristics Nanorods

3.1 Computation of wavenumbers

For analyzing the ultrasonic wave dispersion characteristics in nanorods, we assume that a harmonic type of wave solution for the displacement field $u_(x,t)$ and it can be expressed in complex form as [28, 29],

$$u\left(x,t\right)=\sum_{p=0}^{P-1}\sum_{q=0}^{Q-1}\hat{u}\left(x;\omega_q\right)e^{-j\left(k_p x-\omega_q t\right)}$$

(2)

where, P and Q are the number of time sampling points and number spatial sampling points respectively. ω_q is the circular frequency at the qth time sample. Similarly, k_p is the axial wavenumber at the pth spatial sample point. Substituting Eq. (2) into the governing partial differential equation (Eq. (1)), we get the dispersion relation as follows. Hereafter the subscript p and q in Eq. (2) are dropped for simplified notations. Here $j=\sqrt{-1}$.

The dispersion realtion is: $-k^2 + (e_0 a)^2 \eta^2 \omega^2 k^2 + \eta^2 \omega^2 = 0$, where $\eta = \sqrt{\rho / E}$.

This dispersion relation is solved for the wavenumbers as $k_{1,2} = \pm \eta \omega \sqrt{1-(e_0 a)^2 \eta^2 \omega^2}$.

The wave frequency is a function of wavenumber k, the nonlocal scaling parameter $e_0 a$ and the material properties ($E \& \rho$) of the nanorod. If $e_0 a = 0$, the wavenumber is directly proportional to wave frequency, which will give a non-dispersive wave behavior (for more details refer [29]). The cut-off frequency of this nanorod is obtained by setting $k = 0$ in the dispersion relation. For the present case the cut-off frequency is zero, i.e., the axial wave starts propagating from zero frequency.

3.2 Escape frequency

Fig. 2 shows the spectrum relation plot as a function of nonlocal scale parameter $e_0 a$. From the figure, we see that at certain frequencies, the wavenumber is tending to infinity and this frequency value decreases with increase in the scale parameter. Its value can be analytically determined by looking at the wavenumber expression (Eq. (16)) and setting $k \to \infty$. Which is given as $\omega_{escape} = 1/(e_0 a)\eta$. The escape frequency is inversely proportional to the nonlocal scaling parameter and is independent of the diameter of the nanorod. The variation of the escape frequency with nonlocal scaling parameter is shown in Fig. 3.

3.3 Computation of wave speeds

For the present analysis, we are considering both wave speeds (i.e., phase and group speeds) of the wave and are defined as

$$C_p = \omega/k \text{ and } C_g = \partial \omega / \partial k.$$

These wave speeds are also depend on the nonlocal scaling parameter. When $e_0 a = 0$, both the wave speeds are equal (i.e., $C_p = C_g$), which is already proved for local or classical bars/rods [29]. The phase speed and group speed dispersion curves with wave frequency are shown in Fig. 4 and Fig. 5, respectively.

4 Governing Partial Differential Equations of Nanorod

4.1 Based on second order strain gradient model

Fig. 1 schematically describes a nanobeam under discussion and serves to introduce the axial coordinate x, the axial displacement $u = u(x,t)$, the rod length L, the rod Young's modulus E, and the density ρ. The displacement field and strain for this nanorod are given by $u = u(x,t)$ $\varepsilon = \partial u / \partial x$. First we derive the governing differential equations for the nanorod using second order strain gradient model. The potential (Π) and kinetic (Γ) energies of the nanorod are

$$\Pi^s = 0.5 \int_V \sigma(x)\varepsilon(x)dV = 0.5 \int_V \left(\varepsilon(x) + (e_0 a)^2 \varepsilon^{<ii>} \right)\varepsilon(x)dV \text{ and } \Gamma^s = 0.5 \int_V \rho \left(\dot{u}(x,t) \right)^2 dV.$$

Here superscript s represents for the second order strain gradient model. Assuming a uniform nanorod, these two relations can be rewritten as

$$\Pi^s = 0.5 A \int_V \left(\varepsilon(x) + (e_0 a)^2 \varepsilon^{<ii>}(x) \right)\varepsilon(x)dV \text{ and } \qquad \Gamma^s = 0.5\rho A \int_V \left(\dot{u}(x,t) \right)^2 dV. \text{Applying}$$

Hamilton's principle $\delta \int_{t_1}^{t_2} L^s dt = \delta \int_{t_1}^{t_2} (\Gamma^s - \Pi^s) dt = 0$. Expanding this, and integrating by parts, we obtain the nonlocal governing partial differential equation for the nanorod obtained from the second order strain gradient model as

$$EA(e_0 a)^2 u^{<iv>} + EA u^{<ii>} - \rho A \ddot{u} = 0 \tag{3}$$

This is a fourth order partial differential equation. One can substitute $e_0 a = 0$ in Eq. (3), to recover the local or classical rod equation, which is a second order differential equation.

4.2 Based on fourth order strain gradient model

Now assuming the same displacement field and strain-displacement relations and using the fourth order strain gradient model, the kinetic and potential energies can be expressed as

$$\Pi^f = 0.5 \int_V \left(\varepsilon(x) + (e_0 a)^2 \varepsilon^{<ii>}(x) + (e_0 a)^4 \varepsilon^{<iv>}(x) \right) \varepsilon(x) dV \quad \text{and} \quad \Gamma^f = 0.5 \int_V \rho \left(\dot{u}(x,t) \right)^2 dV.$$

Here superscript f represents the fourth order strain gradient model. For a uniform nanorod, these equations can be rewritten as

$$\Pi^f = 0.5 A \int_0^L \left(\varepsilon(x) + (e_0 a)^2 \varepsilon^{<ii>}(x) + (e_0 a)^4 \varepsilon^{<iv>}(x) \right) \varepsilon(x) dx \text{ and } \Gamma^f = 0.5 \rho A \int_0^L \left(\dot{u}(x,t) \right)^2 dx.$$

Applying the Hamilton's principle, $\delta \int_{t_1}^{t_2} L^f dt = \delta \int_{t_1}^{t_2} (\Gamma^f - \Pi^f) dt = 0$ and integrating by parts, the nonlocal governing differential equation for the nanorod from fourth order strain gradient model is derived as

$$EA(e_0 a)^4 u^{<vi>} + EA(e_0 a)^2 u^{<iv>} + EA u^{<ii>} - \rho A \ddot{u} = 0 \tag{4}$$

This is a sixth order partial differential equation, observe that when $e_0 a = 0$, local/classical rod equation will be recovered.

4.3 Discussion on second order strain gradient model

4.3.1 Analytical solution

Strain gradients can be used to introduce heterogeneity into the continuum. As a result the dispersive character of waves observed in experiments and in the discrete material models can be simulated with a higher accuracy [10, 11]. By homogenising a discrete medium, a second-gradient model can be derived.Analytical solution, which is obtained by combining the constitutive relation with the uniaxial equilibrium equation $\sigma^{<i>} = 0$ (no body forces are considered) and the kinematic relation. The use of Eq. (3) leads to an analytical solution for u of the form $u = C_1 + C_2 x + C_3 \sin(x / e_0 a) + C_4 \cos(x / e_0 a)$ where C_i are constants that have to be determined according to the boundary conditions. The response of the classical continuum is given by the constants C_1 and C_2 only.

4.3.2 Uniqueness

Following [7], the uniqueness of the static analytical solution is investigated next. To this end, it is assumed that two different solutions u_1 and u_2 exist that satisfy the equilibrium equation and the nonhomogeneous boundary conditions. For a proof of uniqueness, the difference between these two solutions $\Delta u = u_1 - u_2$ should vanish. This 'difference solution' should then satisfy the equilibrium equation and the homogeneous boundary conditions. A specimen of length L is considered, and the boundary conditions for the difference solution are taken as $\Delta u = 0$ and $\Delta u^{<i>} = 0$ both at $x = 0$ and at $x = L$. The four boundary conditions lead to the following system of equations: (let $\Omega = L / e_0 a$)

$$\begin{bmatrix} 1 & 0 & 0 & 1 \\ 1 & L & \sin\Omega & \cos\Omega \\ 0 & 1 & 1/L & 0 \\ 0 & 1 & (1/L)\cos\Omega & -(1/L)\sin\Omega \end{bmatrix} \begin{Bmatrix} C_1 \\ C_2 \\ C_3 \\ C_4 \end{Bmatrix} = \begin{Bmatrix} 0 \\ 0 \\ 0 \\ 0 \end{Bmatrix} \tag{5}$$

By eliminating C_1 and C_2, a reduced coefficient matrix for C_3 and C_4 according to Eq. (3) can be determined. For finding a nontrivial solution for

Δu (which corresponds to non-uniqueness) the determinant of this reduced coefficient matrix should vanish, i.e.

$$\begin{vmatrix} \sin\Omega - \Omega & \cos\Omega - 1 \\ \cos\Omega - 1 & -\sin\Omega \end{vmatrix} = 0 \tag{6}$$

that implies $\Omega\sin\Omega + 2\cos\Omega - 2 = 0$, which is satisfied when $\Omega = 2\pi\alpha$ with α an arbitrary integer. Thus, uniqueness cannot be guaranteed for the model of Eq. (3) in the case $L = 2\pi\alpha e_0 a$. In the above procedure, the higher-order boundary conditions are taken as prescribed values for the first derivative of the displacement. The use of different higher order boundary conditions leads to different considerations with respect to uniqueness. Taking prescribed second-order derivatives of the displacement can also lead to non-unique solutions. Although not shown here, in this case the values of $L/e_0 a$ for which non unique solutions are obtained coincide with those obtained via $\Omega\sin\Omega + 2\cos\Omega - 2 = 0$.

4.3.3 Stability

The stability of the second order strain gradient model of Eq. (3) is studied by means of the potential energy density $\Re$, given by $\Re = \int_{\varepsilon} \sigma d\varepsilon$. Substitution of the constitutive equation (3), integrating the higher-order terms by parts, and carrying out the integration results in $\Re = 0.5E(\varepsilon^2 - (e_0 a)^2 \varepsilon^{<i>2})$. In the derivation procedure above, the boundary integrals are assumed to vanish. This has severe implications for the stability of the model: positive terms have a stabilizing effect on the overall response, while negative terms are destabilizing. Thus, the model according to Eq. (3) may become unstable. Finally, the second order strain gradient model can become unstable and uniqueness is not guaranteed. This model bears the closest relation with the discrete model (a positive sign in front of the higher-order term). However, the use of this model in engineering practice is limited due to the possible emergence of non-uniqueness and instability, which may have a devastating effect on its response [30].

5 Ultrasonic Wave Characteristics of Nanorod

5.1 For second order strain gradient model

For analyzing the ultrasonic wave dispersion characteristics in nanorods, we assume that a harmonic type of wave solution for the displacement field $u_{(}x,t)$ and it can be expressed in complex form as $u(x,t) = \hat{u}(x,\omega)e^{-j(kx-\omega t)}$, where $\hat{u}(x,\omega)$ is the frequency domain amplitude of the longitudinal displacement, k is the wavenumber and ω is the angular frequency of the wave motion and $j = \sqrt{-1}$. Substituting this assumed solution in the nonlocal governing equation of nanorod obtained from the second order strain gradient model (Eq. (3)), gives the following dispersion relation,

$(e_0 a)^2 k^4 - k^2 + \eta^2 \omega^2 = 0$, where $\eta = \sqrt{\rho/E}$. This dispersion relation is solved for the wave frequency as $\omega^s = (k/\eta)\sqrt{1 - (e_0 a)^2 k^2}$. The wave frequency is a function of wavenumber k, the nonlocal scaling parameter $e_0 a$ and the material properties $(E \& \rho)$ of the nanorod. For the present analysis, we are considering both wave speeds (phase and group speeds) of the wave and are defined as $C_p^s = \omega^s / k$ and $C_g^s = \partial \omega^s / \partial k$. These wave speeds are also depend on the nonlocal scaling parameter. When $e_0 a = 0$, both the wave speeds are equal (i.e., $C_p^s = C_g^s = 1/\eta$), which is already proved for local or classical bars/rods [29].

5.1.1 Critical Wavenumber

From the dispersion relation obtained from second order strain gradient model, by setting the wave frequency (ω) to zero, gives the critical wavenumber as $k_{cr} = \pm 1/\sqrt{e_0 a}$. The critical wavenumber is purely a function of internal length scale only.

5.1.2 Number of Waves Along the Nanorod

The number of wavelengths (N) along the nanorod is defined as $N = L/\lambda$, where L is the length of the nanorod and λ is the wavelength. The wavelength is related to wavenumber as $\lambda = 2\pi/k$, here the wavenumber is critical. Thus the number of wavelengths along the nanorod is given by $N = L/2\pi\sqrt{e_0 a}$. The variation between the number of waves along the nanorod and the nonlocal scaling parameter will be discussed in next section, for nanorod lengths of $L = 10, 50$ and 100 nm.

5.2 For fourth order strain gradient model

Substituting assumed solution for axial displacement in the nonlocal governing partial differential equation of nanorod obtained from the fourth order strain gradient model (Eq. (4)), gives the following dispersion relation, $-(e_0 a)^4 k^6 + (e_0 a)^2 k^4 - k^2 + \eta^2 \omega^2 = 0$. Solving this dispersion relation for wave frequency as $\omega^f = (k/\eta)\sqrt{1 - (e_0 a)^2 k^2 + (e_0 a)^4 k^4}$. The phase and the group speeds of an ultrasonic wave in nanorod are given as as $C_p^s = \omega^s/k$ and $C_g^s = \partial\omega^s/\partial k$. These wave speeds are also depend on the nonlocal scaling parameter. When $e_0 a = 0$, both the wave speeds are equal (i.e., $C_p^s = C_g^s = 1/\eta$). Also observe that the when $e_0 a = 0$, the wave speeds obtained from second and fourth order strain gradient models are equal to the wave speed in the local or classical rod.

6 Dynamic Response of Nanorods

In this section, the dynamic response of the nanorod is studied. To this end, a semi-infinite nanorod with is considered the left end of which is subjected to an impact excitation, see Fig. 2. The response will be calculated by using the Laplace integral transform over time $u_s(x,s) = \int_0^\infty u(x,t)e^{st}\,dt$. Application of this Laplace transform to the equations of motion Eqs. (3), (4) and the boundary conditions under the assumption that the bar is initially at rest yields the following boundary-value problems:

6.1 *For second order strain gradient model*

$$EA(e_0a)^2 u_s^{<iv>} + EAu_s^{<ii>} + \rho As^2 u_s = 0, \quad x \in [0,\infty[,$$

$$F_0^{second} = EAu_s^{<i>} + EA(e_0a)^2 u_s^{<iii>}, \quad x = 0, \tag{7}$$

$$u_s^{<ii>}, \quad x = 0,$$

6.2 *For fourth order strain gradient model*

$$EA(e_0a)^4 u_s^{<vi>} + EA(e_0a)^2 u_s^{<iv>} + EAu_s^{<ii>} + \rho As^2 u_s = 0, \quad x \in [0,\infty[,$$

$$F_0^{fourth} = EAu_s^{<i>} + EA(e_0a)^4 u_s^{<v>}, \quad x = 0, \tag{8}$$

$$u_s^{<ii>} - (e_0a)^2 u_s^{<iv>} = 0, \quad x = 0,$$

$$u_s^{<iii>}, \quad x = 0,$$

Consider the second order strain gradient model. The general solution to Eq. (7) may be written as $u_s^{second} = A_1 e^{jk_1 x} + A_2 e^{jk_2 x}$, where $k_{1,2}$ are the roots of the characteristic equation $(e_0a)^2 k^4 - k^2 + \eta^2 s^2 = 0$. Since Eq. (7) is a biquadratic equation, $k_{1,2}$ can be found explicitly to give $k_{1,2} = 1/\sqrt{2}(e_0a)\sqrt{1 + \sqrt{1 - 4\eta^2 s^2 (e_0a)^2}}$. The unknown constants A_1, A_2 can be determined by substituting the general solution into the boundary conditions Eqs. (7). This substitution leads to the system of two algebraic equations with respect to A_1 and A_2, which can be solved for A_1 and A_2. Through expressions of A_1 and A_2 we have obtained the solution to the boundary-value problem Eqs. (7), which describes the response of the second-gradient model of the nanorod in the Laplace domain.Consider now the fourth order strain gradient model defined by Eqs. (8). Analogously to the analysis of the second order strain gradient model, we first write the general solution to Eq. (8): $u_s^{fourth} = B_1 e^{jk_1 x} + B_2 e^{jk_2 x} + B_3 e^{jk_3 x}$, where $k_{1,2,3}$ are the roots of the characteristic equation. $-(e_0a)^4 k^6 + (e_0a)^2 k^4 - k^2 + \eta^2 s^2 = 0$, Substitution of the general solution into the boundary conditions Eqs. (8) yields a system of three algebraic equations with respect to B_1, B_2 and B_3. Such system can be easily solved to give expressions for $B_{1,2,3}$ as functions of $k_{1,2,3}$, s, and the other system parameters. Therefore, taking into account that $k_{1,2,3}$ can be found either analytically or by means of any standard program for finding the roots of a polynomial, we may claim that the impact response of the fourth order strain gradient model of the nanorod in the Laplace domain has been found.

To inverse solutions Eqs. (7) and (8) into the time domain, the following integral has to be calculated $u(x,t) = (1/2\pi j) \int_{\gamma - j\infty}^{\gamma + j\infty} u_s(x,s)e^{-st} ds$, where γ is a positive constant that has to be chosen larger than the real part of all singularities of the function $u_s(x,s)$. Since neither $u_s^{second}(x,s)$ nor $u_s^{fourth}(x,s)$ have singularities at the right half-plane of the complex s-plane, we may choose any positive value for γ. The numerical inversion of the solutions Eqs. (7) and (50) will depict the dynamic response of the nanorods. One can easily obtain the dynamic response of the nanorods in terms of the displacement (u), strain $(u_x = u^{<i>})$, velocity $(\dot{u} = \partial u / \partial t)$, etc, from Eqs. (7) and (8).

In dynamics, the differences between the models become more pronounced. The second order strain gradient model fails completely due to imaginary propagation velocities of the higher wave numbers (will be discussed more detail in the next section). These higher wave numbers are especially dominant when shock waves are generated. Thus, this second order strain gradient model is not suitable to study practical dynamic problems. On the other hand, in the fourth order strain gradient model, all propagation velocities are real, and realistic responses can be found. The model shows a dispersive character, and the intensity of the dispersion depends on the internal length scale parameter. However, waves are found that propagate with an unrealistically high velocity, but this only concerns the higher wave numbers, and their influence on the global response appears to be limited in the presented analysis.

7 Nonlocal Governing Partial Differential Equation for Nanorods

Fig. 12 schematically describes a nanorod under discussion and serves to introduce the axial coordinate x, lateral coordinate y, the axial displacement $u = u(x,t)$, the Young's modulus E, the density ρ, the Poisson's ratio v and cross sectional area A. The displacement field (X-direction), strain, strain rate and particle velocity associated with the displacement field in X-direction for this nanorod are given by $u = u(x,t) \rightarrow \varepsilon_{xx} = u^{<i>} \rightarrow \dot{\varepsilon}_{xx} = \dot{u}^{<i>}$ and $V = \dot{u}$.

Due to Poisson's ratio v, there are displacement fields v and w in $Y-$ and $Z-$ directions, respectively. For example the strain and the derivative of the displacement with time in the $Y-$direction are, respectively, $\varepsilon_{yy} = -v\varepsilon_{xx}$ and $\dot{v} = -vy\dot{\varepsilon}_{xx}$. The kinetic energy of the infinitesimal length Δx of the rod is $\Delta \Pi^e = 0.5\rho A \Delta x(V^2 + v^2\zeta^2\dot{\varepsilon}_{xx}^2)$, where ρ is the density, A is the cross-sectional area and ζ is the radius of gyration of the solid circular cross-section. Here, an effective density is used to incorporate the effect of lateral inertia in a one-dimensional nonlocal wave equation. An effective density ρ_{eff} is now introduced such that the kinetic energy of the element Δx is $\Delta \Pi^e_{eff}$, where $\Delta \Pi^e_{eff} = 0.5\rho_{eff} A \Delta x V^2$. Using Newton's second law, the net longitudinal force acting on the element Δx (see Fig. 12) is $\Delta \sigma_{xx} A = \rho_{eff} A \ddot{u} \Delta x$.

In the limiting case as $\Delta x \rightarrow 0$, equilibrium eqn. can be written as $\sigma_{xx}^{<i>} = \rho_{eff}\ddot{u}$. To find the relationship between ρ_{eff} and ρ, a functional $\bigcup_e$ is defined as the kinetic energy error when using the effective density, i.e.,

$$\bigcup_e = \iint \left[\Delta \Pi^e - \Delta \Pi^e_{eff} \right] dx\, dt = 0.5 \iint \left[\rho A\left(V^2 + v^2\zeta^2\dot{\varepsilon}_{xx}^2\right) - \left(\rho_{eff} A V^2\right) \right] dx\, dt.$$

Substituting for strain rate and particle velocity in above relation, the minimization of $\bigcup_e$ requires minimization of the following functional:

$$\bigcap_e = \iint \left\{ \left[\rho V^2 + \rho v^2\zeta^2 \left(\partial V / \partial x\right)^2 - \rho_{eff} V^2 \right] \right\} dx\, dt.$$ The functional can be written interms of the displacement u as $\bigcap_e = \iint \left\{ \left(\rho - \rho_{eff}\right)\dot{u}^2 + \rho v^2\zeta^2 \left[\dot{u}^{<i>} \right]^2 \right\} dx\, dt$. The integrand of the functional $\bigcap_e$ is $f\left(\dot{u}, \dot{u}^{<i>}\right) = \left(\rho - \rho_{eff}\right)\dot{u}^2 + \rho v^2\zeta^2 \left[\dot{u}^{<i>} \right]^2$. In order

to minimize $\bigcap_e$, the integrand $f(\dot{u}, \partial \dot{u}/\partial x)$ should satisfy the following equation:

$$\partial(\dot{u}^{<i>})\left(\partial f\left(\dot{u}, \dot{u}^{<i>}\right)/\partial(\dot{u}^{<i>})\right) - \partial/\partial t\left[\partial f\left(\dot{u}, \dot{u}^{<i>}\right)/\partial(\dot{u}^{<i>})\right] = 0 \tag{9}$$

Therefore, Eq. (9) is identical to the following partial differential equation: $\rho_{eff}\ddot{u} = \rho\ddot{u} - \rho v^2 \zeta^2 \partial^2 \ddot{u}^{<ii>}$. This relation gives the relationship between the effective density and the density that minimizes the effective density error, $\bigcup_e$ over the length of the nanorod and the time of motion. Substituting this expression in equilibrium equation gives $\sigma_{xx}^{<i>} = \rho\ddot{u} - \rho v^2 \zeta^2 \ddot{u}^{<ii>}$. The constitutive model employed here is that obtained from the theory of nonlocal/nonclassical continuum mechanics. For thin rods the nonlocal constitutive realtion can be written in the following one dimensional form $\sigma_{xx} - \alpha^2 \sigma_{xx}^{<ii>} = E\varepsilon_{xx} = Eu^{<i>}$, where E is the modulus of elasticity, σ_{xx} and ε_{xx} are the local stress and strain components in the x direction, respectively and $\alpha = e_0 a$, nonlocal scaling parameter. Differentiating the nonlocal constitutiv relation mentione above, with respect to x on both sides, gives $\sigma_{xx}^{<ii>} - \alpha^2 \sigma_{xx}^{<iii>} = E\varepsilon_{xx}^{<i>} = Eu^{<ii>}$. Substituting these in obtained equilibrium realtions leads to

$$Eu^{<ii>} = \rho\ddot{u} - \rho v^2 \zeta^2 \ddot{u}^{<ii>} - \alpha^2 \rho\ddot{u}^{<ii>} + \alpha^2 \rho v^2 \zeta^2 \ddot{u}^{<vi>} \tag{10}$$

Eq. (10) is the consistent fundamental governing equation of motion for nonlocal rod model including the effect of lateral inertia/Poisson's effect. When $\alpha = e_0 a = 0$ and $v = 0$, it is reduced to the equation of local or classical rod model.

8 Numerical Experiments, Results and Discussion

For the present analysis, a single walled carbon nanotube is assumed as a nanorod. The values of the radius, thickness, Young's modulus and density are assumed as 3.5 nm, 0.35 nm , 1.03 TPa , and, 2300 kg/m³ respectively.

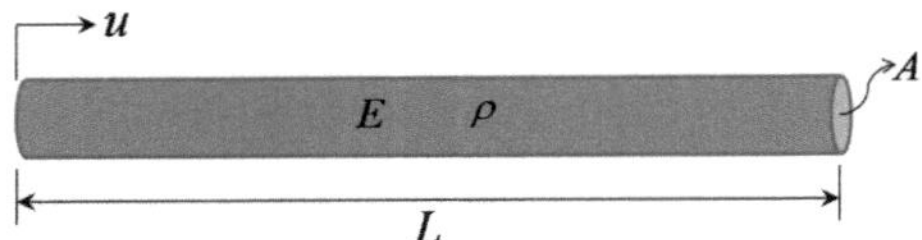

Figure 1: A nanorod, showing length L, Young's modulus E, density ρ, cross-sectional area A and longitudinal displacement u.

8.1 Stress gradeint model

Fig. 2 shows the real and imaginary parts of the axial wavenumber of a nanorod. The thick lines represent the real part and the thin lines show the imaginary part of the wavenumbers. From Fig. 2, for a nanorod, it can be seen that there is only one mode of wave propagation i.e., axial or longitudinal. For local or classical model, the wavenumbers for the axial mode has a linear variation with the frequency which is in the THz range. The linear variation of the wavenumbers denote that the waves will propagate non-dispersively, i.e, the waves do not change their shapes as they propagate. On the other hand, the wavenumbers obtained from nonlocal elasticity have a non-linear variation with the frequency, which indicates that the waves are dispersive in nature. However, the wavenumbers of this wave mode have a substantial real part starting from the zero frequency. This implies that the mode starts propagating at any excitation frequency and does not have a cut-off frequency. At certain frequency (Escape frequency), the wavenumbers tending to infinite as shown in Fig. 2. So, the nonlocal elasticity shows that the wave will propagate up to certain frequencies only and after that the wave will not propagate.

The escape frequencies are purely a function of the nonlocal scaling parameter. The variation of the escape frequency with nonlocal parameter is shown in Fig. 3. It shows that, as $e_0 a$ increases the escape frequency decreases, such variation can also observed from Fig. 2. For very small values of $e_0 a$, the escape frequencies are very large, and at higher values of $e_0 a$ the escape frequencies are very small and approach to a constant value.

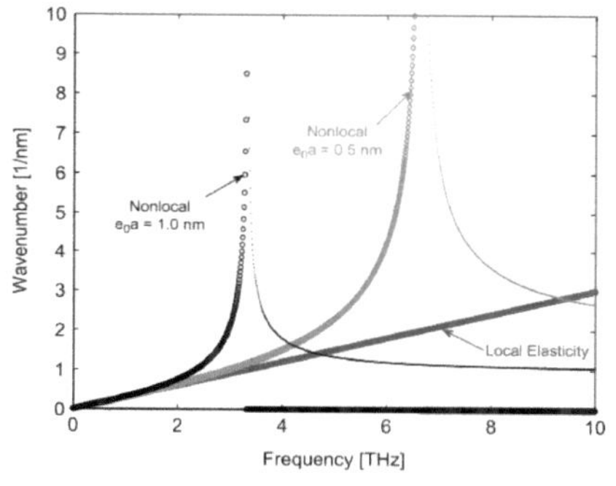

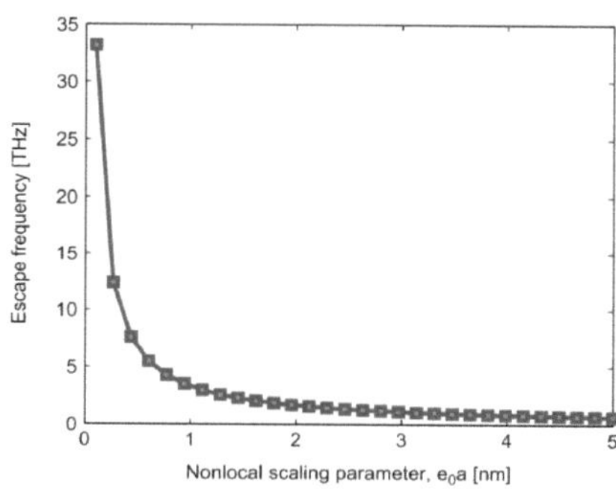

Figure 2: A Comparison of the wavenumber (thick lines-real; thin lines-imaginary) dispersion in a nanorod obtained from local and nonlocal elasticity theories.

Figure 3: Escape frequency variation of the axial wave in a nanorod with nonlocalscaling parameter.

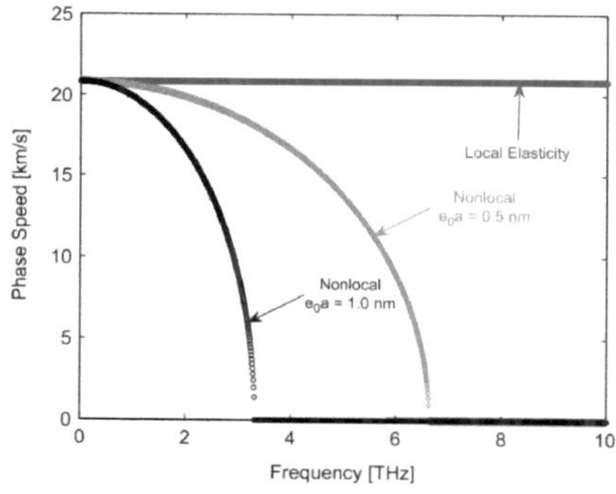

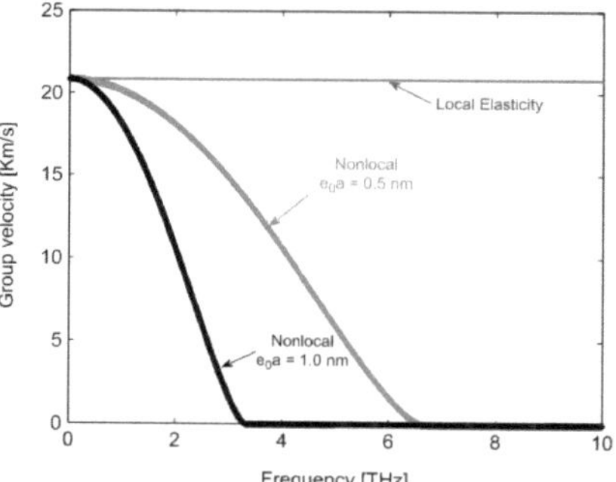

Figure 4: A comparison of the phase speed dispersion in a nanorod obtained from local and nonlocal elasticity theories.

Figure 5: A Comparison of the group speed dispersion in a nanorod obtained fromlocal and nonlocal elasticity theories.

Fig. 4 and Fig 5, plots the wave speeds for the nanorod obtained from both local and nonlocal models. Because of the linear variation of wavenumber with wave frequency from local elasticity, the phase speed $(C_p = Re(\omega/k(\omega)))$ and group speed $(C_g = Re(d\omega/dk(\omega)))$ for the axial mode has a constant value for all the frequencies and hence, the wave does not change its shape as it propagate. It can also be observed, that the axial wave speed is similar for local and nonlocal cases at zero frequency. In nonlocal elasticity, the wave number tends to infinite at certain frequencies, so that the phase and group speeds are tending to zero at those frequencies, indicating localization and stationary behavior.

It can be concluded that the wave dispersion characteristics in a nanorod is drastically different for local and nonlocal models. Where local model predicts that he wave will propagate at all frequencies, but the nonlocal model shows that the wave will propagate up to certain frequencies only depending on the nonlocal scaling parameter. The results presented in this paper can provide useful guidance for the study and design of the next generation of nanodevices that make use of the wave propagation properties of carbon nanotubes.

8.2 Strain gradient models

The spectrum and dispersion curves for nanorod obtained from both strain gradient models (assuming e_0a = 0.02 nm as length increases $e_0a/L \ll 1$) are shown in Fig. 6. The results obtained from classical continuum model are also shown in those figures for comparison.

The spectrum and dispersion relations obtained from the classical continuum model shows that the waves in nanorod are non-dispersive i.e., the wavenumber has a linear relation with the wave frequency or the phase or group speeds are constant (see Figs. 6 - 9). Both the strain gradient models shows that the waves in nanorod are dispersive in nature. It can also be seen that the fourth-order strain gradient model give improved approximation over the second order strain gradient model, as compared to the classical continuum model. And also the results are compared with the Born-Karman model [14], and the stress gradient model [18].

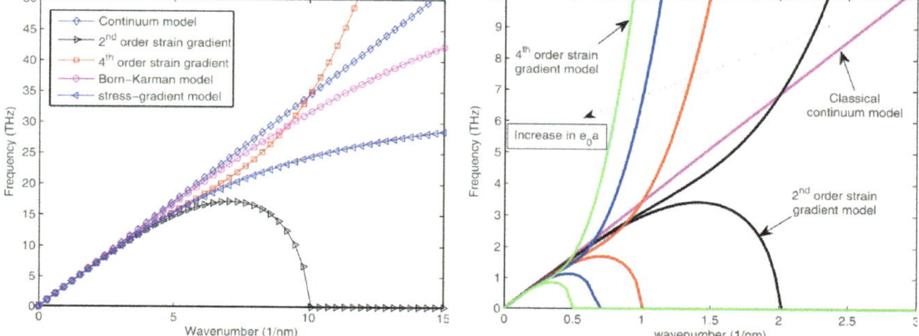

Figure 6: Wave number dispersion with wave frequency obtained from classical continuum model, second and fourth order strain gradient models.

Figure 7: Effect of nonlocal scaling parameter on wavenumber dispersion in nanorod.

The instability of the second order strain gradient model also becomes manifest in Fig. 6 for wave numbers larger than $1/\sqrt{e_0 a}$ the angular frequency and the phase velocity become imaginary. This means that waves with larger wave numbers (or, equivalently, with smaller wave lengths) cannot propagate through this medium. Instead, the imaginary frequency and velocity imply that the response occurs everywhere in the medium instantaneously. This is physically unrealistic. Therefore, these smaller wave lengths should not be considered. Filtering shorter waves occurs automatically in a discrete medium, where wave lengths smaller than two times the particle size cannot be monitored. However, in a continuous medium, all wave lengths can in principle be present. Especially when shock waves are investigated, all wave lengths are triggered by the loading. The imaginary angular frequency (or phase velocity) of these high-frequent waves prohibits a proper wave propagation simulation with this model. The cut-off value for the wave number, i.e. the wave number for which the angular frequency is zero, dominates the static response of the second order strain gradient gradient model. This cut-off value emerges at $k = 1/\sqrt{e_0 a}$, see Fig. 3. While for the fourth-gradient model it only concerns the higher wave numbers, see Fig. 3. However, in the response of the fourth order strain gradient model the effect of these high frequency waves are of minor importance. Due to the non-uniqueness and instability of the second order strain gradient model will give quite different wave behavior compare to the fourth order strain gradient model.

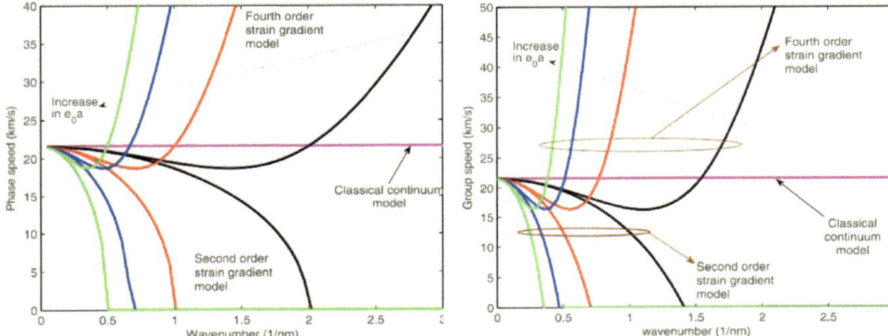

Figure 8: Effect of nonlocal scaling parameter on phase speed dispersion in nanorod.

Figure 9: Effect of nonlocal scaling parameter on Group dispersion in nanorod.

The effect of $e_0 a$ on the wave behavior in nanorods is shown in Figs. 7 to 9. Fig. 7 shows the effect of $e_0 a$ on wavenumber dispersion in nanorod obtained from different models. As $e_0 a$ increase, the wavenumbers obtained from both strain gradient models will decrease with wave frequency. And the corresponding wave speeds will decrease (as $e_0 a$ increases) as shown in Figs. 8 and 9. The phase speeds obtained from the second order strain gradient model are zero at a particular wave frequency (i.e., at critical wavenumber) and this critical wavenumber is also decreases with as $e_0 a$ increases (see Fig. 8). A similar type of phenomena is also observed for group speed variation as shown in Fig. 9.

As shown in Fig. 6, for the case of the second order strain gradient model, the wave frequency is zero at zero wavenumber and also at the value of 2.02 nm^{-1}. The wavenumber at which the wave frequency is zero is called as critical wavenumber k_{cr} (see Eq. (33)). This critical wavenumber is purely a function of the nonlocal scaling parameter. As e_0a increases, the critical wavenumber decreases as shown in Fig. 10. It is found that, for second order strain gradient model, for wavenumbers higher than k_{cr}, the waves cannot not propagate through this medium.

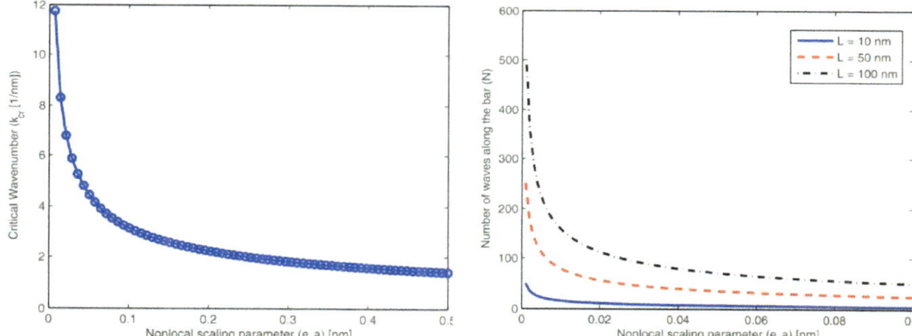

Figure 10: Critical wave number obtained from second orderstrain gradient model variation with nonlocal scaling parameter.

Figure 11: Relation between the number of waves along thenanorod and the nonlocal scaling parameter obtained from second order strain gradient model.

The effect of the nonlocal scaling parameter on the number of waves (N) along the nanorod is shown in Fig. 11, for rod lengths of 10, 50 and 100 nm. A relation among the number of waves, nonlocal scaling parameter and the length of the rod is presented in previous section. This relation is obtained from the second order strain gradient model only. For number of waves along the nanorod are directly proportional to the length of the nanorod. The number of waves along the nanorod will decrease as e_0a increases (see Fig. 11). For very small values of e_0a, the the number of waves along the nanorod are very high, for long length nanorods. For $e_0a > 0.1\,nm$, the number of waves along the rod are almost constant and this constant number is different for different lengths of nanorod.

Finally, the wave dispersion behavior of nanorods, incorporating a constitutive law that includes a length scale, can be analyzed using a direct Newtonian approach (stress gradient model) and a variational approach (strain gradient model). The latter approach provides the governing field equations and the variationally consistent sets of boundary conditions. Incorporation of small length scale effects are found to significantly affect the wave dispersion behavior in nanorods. Strain-gradient elasticity is quite useful even in materials exhibiting small non-local characteristic length scales e.g. in analysis of defects.

8.3 Nonlocal stress gradient model with lateral inertia

Fig. 13 shows the variation of axial wavenumber of a nanorod with wave frequency. From Fig. 13, for a nanorod, it can be seen that there is only one mode of wave propagation i.e., axial or longitudinal. For local or classical model, the wavenumbers for the axial mode has a linear variation with the frequency which is in the tera-hertz (THz) range. The linear variation of the wavenumbers denote that the waves will propagate non-dispersively, i.e, the waves do not change their shapes as they propagate. On the other hand, the wavenumbers obtained from strain gradient/nonlocal stress gradient models have a non-linear variation with the frequency, which indicates that the waves are dispersive in nature. However, the wavenumbers of this wave mode have a substantial real part starting from the zero frequency. This implies that the mode starts propagating at any excitation frequency and does not have a cut-off frequency (refer Fig. 13).

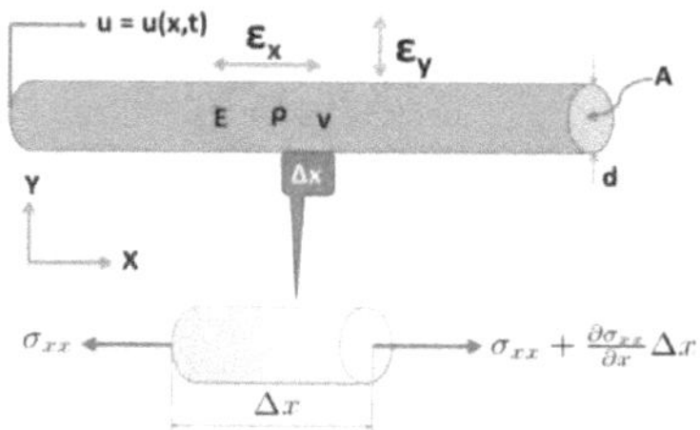

Figure 12: A nanorod, showing Young's modulus E, density ρ, Poisson's ratio v, diameter d, cross-sectional area A, longitudinal displacement u(x,t), strain along X-direction e_x and strains along Y-direction e_y .

Born-Karman model (Based on Lattice Dynamics):

Dispersion relation:
$$\omega = \frac{2}{a}\sqrt{\frac{E}{\rho}}\sin\left(\frac{k \times a}{2}\right).$$

The spectrum curves for nanorod obtained from both second and fourth order strain gradient models are also shown in Fig. 13, for comparison. As said, the spectrum relations obtained from the classical continuum model shows that the waves in nanorod are nondispersive. But both the strain gradient models shows that the waves in nanorod are dispersive in nature. It can also be seen that the fourth order strain gradient model give improved approximation over the second order strain gradient model, as compared to the classical continuum model (such observations are also made in ref. [45]). And also the results are compared with the Born-Karman model, as well as the nonlocal stress gradient model. The present result shows an improved approximation over the second order strain gradient model, see Fig. 13. It can be concluded that, the instability of the second order strain gradient model can be overcome by considering the inertia gradients.

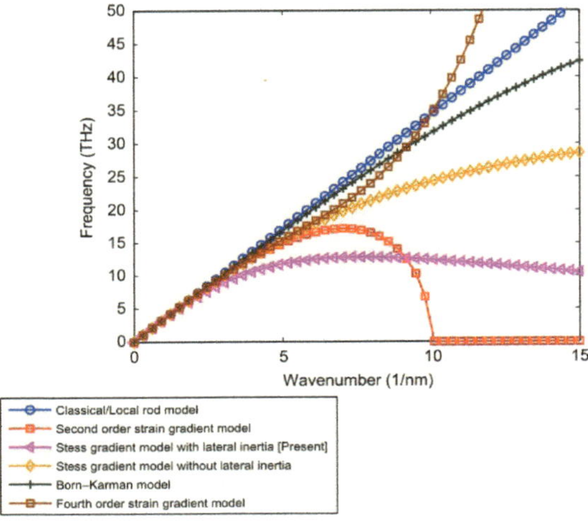

Figure 13: A Comparison of the wavenumber dispersion with wave frequency in a nanorod obtained from classical/local and nonlocal theories suggested in literature with the present result, to show the effect of lateral inertia.

The instability of the second order strain gradient model also becomes manifest in Fig. 13, for wave numbers larger than $\alpha^{-0.5}$ the angular frequency and velocity become zero or imaginary (for more details ref. [45]). This means that waves with larger wave numbers (or, equivalently, with smaller wavelengths) cannot propagate through this medium. Instead, the imaginary frequency and velocity imply that the response occurs everywhere in the medium instantaneously. This is physically unrealistic. Therefore, these smaller wavelengths should not be considered. Filtering shorter waves occurs automatically in a discrete medium, where wavelengths smaller than two times the particle size cannot be monitored. However, in a continuous medium, all wavelengths can in principle be present. Especially when shock waves are investigated, all wavelengths are triggered by the loading. The imaginary angular frequency (or phase velocity) of these high-frequent waves prohibits a proper wave propagation simulation with this model. The cut-off value for the wave number, i.e., the wave number for which the angular frequency is zero, dominates the static response of the second order strain gradient model. This cut-off value emerges at $k = \alpha^{-0.5}$, see Fig. 13. While for the fourth-gradient model it only concerns the higher wave numbers, see Fig. 13. However, in the response of the fourth order strain gradient model the effect of these high frequency waves are of minor importance. Due to the nonuniqueness and instability of the second order strain gradient model will give quite different wave behavior compare to the fourth order strain gradient model. The unstable second order strain gradient model can made stable by considering the inertia gradients in the formulation.

The effect of the scaling parameter on the wave dispersion relation in nanorod is shown in Fig. 14. As the nonlocal scaling parameter increases, the frequency of the wave will decrease with an in crease in wavenumber. Here α values are assumed from 0.0 to 1.0 nm. As α tends to larger value, the wave frequency becomes very smaller at higher values of wavenumber as shown in Fig. 14. It can be seen that, while dealing with the nonlocal elasticity theory including the effect of lateral inertia, on should not neglect the scaling parameter. So, the scaling parameter play an important role while dealing with the dynamics of nanostructures.

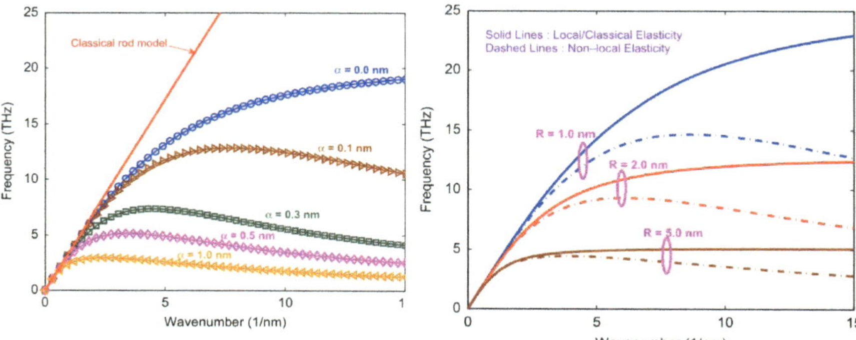

Figure 14: Wavenumber dispersion (including the effect of lateral inertia) with wave frequency in a nanorod for various values of the nonlocal scaling parameter.

Figure 15: Wavenumber dispersion (including the effect of lateral inertia) with wave frequency in a nanorod obtained from both the local and nonlocal formulations for different radii of nanorods.

The effect of radius of nanorod, on the wave dispersion relation based on the present formulation is shown in Fig. 15. For the present analysis, we considered three different radii of nanorods i.e., 1.0 nm, 2.0 nm and 5.0 nm. For comparison, the results obtained from local/classical elasticity are also shown in the figure. As the radius of the nanorod increases, the wave frequency becomes almost constant for wavenumbers larger than 5.0 nm^{-1}. The effect of the lateral inertia shows that, as the radius of the nanorod increases, the wavenumber decreases and will not become constant irrespective of the wavenumber. the instability of the second order strain gradient model can be overcome by considering the inertia gradients.

It can be concluded that the wave dispersion characteristics in a nanorod is drastically different for local and nonlocal models. Where local model predicts that the wave will propagate at all frequencies, but the nonlocal model shows that the wave will propagate up to certain frequencies only depending on the nonlocal scaling parameter. It has also been shown that, the unstable second order strain gradient model can be replaced by considering the inertia gradient terms in the formulations. The results presented in this paper can provide useful guidance for the study and design of the next generation of nanodevices that make use of the wave propagation properties of carbon nanotubes.

9 Concluding Remarks

The nonlocal stress gradient theory has been incorporated into classical rod model to capture unique features of the nanorods (ultrathin structures). The strong effect of the nonlocal scale has been obtained which leads to substantially different wave behaviors of nanorods from those of classical rods. Nonlocal bar model is developed for nanorods. Explicit expressions are derived for wavenumbers and wave speeds of nanorods. The studies also shows that the nonlocal scale parameter introduces certain band gap region in axial wave mode where no wave propagation occurs. This is manifested in the spectrum cures as the region where the wavenumber tends to infinite (or wave speed tends to zero).

Next, the ultrasonic wave dispersion characteristics of a nanorod are analyzed using nonlocal second and fourth order strain gradient models. It has been shown that the second order strain gradient model can become unstable and uniqueness is not guaranteed. The wave analysis in nanorod shows that the fourth order strain gradient model gives approximate results over the second order strain gradient model for dynamic analysis. The second order strain gradient model gives a critical wavenumber at certain wave frequency, where the wave speed is zero. The ultrasonic wave characteristics of the nanorod obtained from the nonlocal strain gradient models are compared with the classical continuum model. A relation among the number of waves along the nanorod, the nonlocal scaling parameter and the length of the nanorod is obtained from the nonlocal second order strain gradient model. The effect of nonlocal scaling parameter and the length of the nanorod on the number of waves along the nanorod are also captured in this work. Finally, the dynamic response behavior of nanorods is explained from both the strain gradient models.

Finally, the nonlocal stress gradeint theory has been incorporated into classical rod model by considering the lateral inertia effect to capture unique features of the nanorods. It has been shown that, the unstable second order strain gradient model presented in literature can be made stable by considering the inertia gradient terms in the formulations. The effect of the nonlocal small scale parameter and the size of the nanorod on the wavenumber dispersion relation are also investigated in the present manuscript.

The results can provide useful guidance for the study and design of the next generation of nanodevices that make use of the axial wave propagation properties of nanorods.

References

[1]   H. W. Kroto, J. R. Heath, S. C. Obrien, R. F. Curl, and R. E. Smalley, C-60 - Buckminsterfullerene, *Nature*, 318 (1985) 162 - 163.

[2]   B. O. Dabbousi, J. RodriguezViejo, F. V. Mikulec, J. R. Heine, H. Mattoussi, R. Ober, K. F. Jensen, and M. G. Bawendi, (CdSe)ZnS core-shell quantum dots: Synthesis and characterization of a size series of highly luminescent nanocrystallites, *Journal of Physical Chemistry B*, 101 (1997) 9463 - 9475.

[3]   J. L. Vossen and W. Kern, Thin Film Processes, *Academic Press, Inc.*, London, UK, (1978).

[4]   S. D. Senturia, Microsystem Design, *Kluwer Academic Publishers*, Norwell, (2001).

[5]   C. R. Martin, Membrane-based synthesis of nanomaterials, *Chemistry of Materials*, 8 (1996) 1739 - 746.

[6]   H. Askes, A. S. J. Suiker, L. J. Sluys, A classification of higher order strain-gradient models-linear analysis, *Archeive of applied mechanics*, 72 (2002) 171 - 188.

[7]   B. S. Altan, E. C. Aifantis, On some aspects in the special theory of gradient elasticity, *J Mech Behavior Mat*, 8 (1997) 231 - 282.

[8]   L. J. Sluys, Wave propagation, localisation and dispersion in softening solids, Dissertation, Delft University of Technology, (1992).

[9]   H. B. Muhlhaus, E. C. Aifantis, A variational principle for gradient plasticity, *Int J Solids Struct*, 28 (1991) 845 - 857.

[10]   C. S. Chang, J. Gao, Second-gradient constitutive theory for granular material with random packing structure, *Int J Solids Struct*, 32 (1995) 2279 - 2293.

[11]   H. B. Muhlhaus, F. Oka, Dispersion and wave propagation in discrete and continuous models for granular materials, *Int J Solids Struct*, 33 (1996) 2841 - 2858.

[12]   N. Triantafyllidis, S. Bardenhagen, On higher gradient continuum theories in 1-D non-linear elasticity. Derivation from and comparison to corresponding discrete models, *J Elas.*, 33 (1993) 259 - 293.

[13]   C. A. Eringen, D.G.B. Edelen, On non-local elasticity, *International journal of Engineering Sciences* 10 (1972) 233.

[14]   C. Eringen, On differential equations of nonlocal elasticity and solutions of screw dislocation and surface waves, *Journal of Applied Physics* 54 (1983) 4703.

[15]   C. Eringen, Linear theory of non-local elasticity and dispersion of plane waves, *International journal of Engineering Sciences*, 10 (1972) 425.

[16]   C. Eringen, Non-local Polar Field Models, *Academic, New York*, (1996).

[17]   S. Narendar, S. Gopalakrishnan, Terahertz wave characteristics of a single-walled carbon nanotube containing a fluid flow using the nonlocal Timoshenko beam model, *Physica E: Low-dimensional Systems and Nanostructures*, 42 (2010) 1706.

[18]   S. Narendar, S. Gopalakrishnan, Nonlocal scale effects on ultrasonic wave characteristics of nanorods, *Physica E: Low-dimensional Systems and Nanostructures*, 42 (2010) 1601-1604.

[19]   S. Narendar, S. Gopalakrishnan, Investigation of the effect of nonlocal scale on ultrasonic wave dispersion characteristics of a monolayer graphene, *Computational Materials Science*, 49 (2010) 734-742.

[20]   S. Narendar, S. Gopalakrishnan, Ultrasonic wave characteristics of nanorods via nonlocal strain gradient models, *Journal of Applied Physics*, 107 (2010) 084312.

[21]   S. Narendar, S. Gopalakrishnan, Strong nonlocalization induced by small scale parameter on terahertz flexural wave dispersion characteristics of a monolayer graphene, *Physica E: Low-dimensional Systems and Nanostructures*, 43 (2010) 423-430.

[22]   Y. Chen, J. D. Lee, A. Eskandarian, Atomistic viewpoint of the applicability of microcontinuum theories, *Int. J Solids Struct.*, 41 (2004) 2085.

[23]   J. Peddieson, G. R. Buchanan, and R. P. McNitt, Application of nonlocal continuum models to nanotechnology, *International Journal of Engineering Science*, 41 (2003) 305 - 312.

[24]   Q. Wang and C. M. Wang, On constitutive relation and small scale parameter of nonlocal continuum mechanics for modeling carbon nanotubes, *Nanotechnology*, 18 (2007), 075702 1 - 4.

[25]   L. J. Sudak, Column buckling of multiwalled carbon nanotubes using nonlocal continuum mechanics. *Journal of Applied Physics* 94 (2003) 7281 - 7287.

[26]   Y. Q. Zhang, G. R. Liu, X. Y. Xie, Free transverse vibration of double-walled carbon nanotubes using a theory of nonlocal elasticity, *Physical Review B* B 71 (2005) 195404.

[27]   S. Narendar, S. Gopalakrishnan, Nonlocal scale effects on wave propagation in multi-walled carbon nanotubes, *Computational Materials Science*, 47 (2009) 526 - 538.

[28]   S. Gopalakrishnan, A. Chakraborty, D. Roy Mahapatra, Spectral finite element method, *Springer Verlag London Ltd.* (2008).

[29]   J. F. Doyle, Wave propagation in structures, *Springer-Verlag Inc*, New York, 1997.

[30]   H. Askes, A. S. J. Suiker, L. J. Sluys, Dispersion analysis and element-free Galerkin simulations of higher-order strain gradient models, *Mat Phys Mech*, 3 (2001) 12 - 20.